Life Under the Sea

Corals

by Cari Meister

Bullfrog Books

Ideas for Parents and Teachers

Bullfrog Books let children practice reading informational text at the earliest reading levels. Repetition, familiar words, and photo labels support early readers.

Before Reading
- Ask the child to think about coral. Ask: What do you know about corals?
- Look at the picture glossary together. Read and discuss the words.

Read the Book
- Read the book to the child, or have him or her read independently.

After Reading
- Prompt the child to think more. Ask: Is a coral an animal or a plant? What would it be like to live in one spot your whole life?

Bullfrog Books are published by Jump!
5357 Penn Avenue South
Minneapolis, MN 55419
www.jumplibrary.com

Library of Congress Cataloging-in-Publication Data
Meister, Cari.
  Corals / by Cari Meister.
    p. cm. -- (Bullfrog books. Life under the sea)
  Summary: "This photo-illustrated book for early readers tells the story of how a coral grows, finds food, and becomes part of a reef"-- Provided by publisher.
  Audience: K to grade 3.
  Includes bibliographical references and index.
  ISBN 978-1-62031-031-1 (hardcover : alk. paper) --
ISBN 978-1-62496-049-9 (ebook)
  1. Corals--Juvenile literature. 2. Coral reefs and islands--Juvenile literature. I. Title.
  QL377.C5M45 2014
  593.6--dc23   2013001957

Series Editor   Rebecca Glaser
Book Designer   Ellen Huber
Photo Researcher   Rebecca Pettiford

Photo Credits: All photos by Shutterstock except: Corbis, 14-15; iStockPhoto, 3b, 24; ScienceSource, 17, 23bl

Printed in the United States of America at Corporate Graphics, North Mankato, Minnesota.
5-2013 / PO 1003

10 9 8 7 6 5 4 3 2 1

Table of Contents

A City in the Sea

Look at the reef.

It goes for miles.
It is like a city.

eel

Fish swim.
Sea horses hide.
Eels hunt.

The reef is their home.

But what
makes
the reef?
Corals!

A coral is a small animal.

It has a soft body.

11

A coral has a hard shell.
When it dies, its shell stays.

The reef grows.

A coral stays in one spot.
How does it get food?

octopus

tentacle

Look at its tentacles. They wave.
When food floats by, they sting it.

What's for lunch?

plankton

Plankton!
They are
tiny plants
and animals.

17

What eats coral?
A parrotfish does.
Turtles do too.

There are all kinds of corals.

Some look like flowers.
Some look like stars.
Some look like brains!

Parts of a Coral

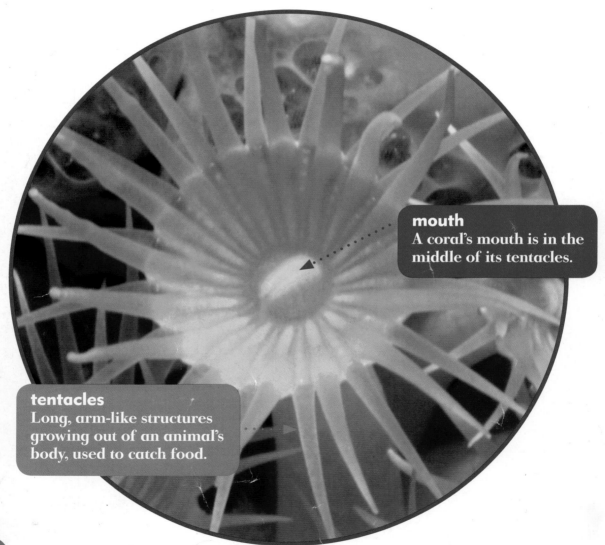

mouth
A coral's mouth is in the middle of its tentacles.

tentacles
Long, arm-like structures growing out of an animal's body, used to catch food.

Picture Glossary

parrotfish
A large colorful fish that has a beak like a parrot.

reef
A ridge of rocks, sand, or coral that is close to the surface of an ocean.

plankton
Tiny plants and animals that float in the sea.

shell
A hard protective covering.

23

Index

To Learn More

Learning more is as easy as 1, 2, 3.

1) Go to www.factsurfer.com

2) Enter "coral" into the search box.

3) Click the "Surf" button to see a list of websites.

With factsurfer.com, finding more information is just a click away.